STILL COMING HOME

Denver Veterans Writing

edited by
Jason Arment, Steven Dunn,
and Bethany Strout

Still Coming Home

v3.0

The opinions expressed in this manuscript are solely the opinions of the author and do not represent the opinions or thoughts of the publisher. The author has represented and warranted full ownership and/or legal right to publish all the materials in this book.

ISBN: 978-0-9643560-1-6

Colorado Humanities & Center for the Book
7935 E. Prentice Ave., Suite 450
Greenwood Village, CO 80111
coloradohumanities.org All rights reserved - used with permission.

PRINTED IN THE UNITED STATES OF AMERICA

think. learn.
join the conversation.

Dear Reader,

Thank you from all of us who contributed to and edited this anthology. We appreciate your engagement with this compilation of work from veterans and people related to veterans. What you will see here are essays, stories, and poems that are as complicated and diverse as the individuals who make up the military.

This project was born out of a workshop led by volunteer veterans who are also writers. Some veterans wanted to write but didn't know how to get started. Some veterans were afraid people wouldn't understand their truths. Some veterans wanted to push their current writing process further.

Some have kept journals for 50 years. Some are women who've endured misogyny on top of the standard military stressors. Some are Mexican and Native American and have seen their ancestors' stories erased from the dominant military narratives in the United States. But what they all have in common is a sincere effort to tell their stories in the most vulnerable and honest way that they could. We are fortunate to be able to offer this collection to you on behalf of the authors.

Thank you again, Reader, for your time, ears, and heart.

With love,
The Editors: Jason Arment, Steven Dunn, and Bethany Strout

Contents

ESSAYS

POEMS

FICTION

Essays

CATH COLWELL

Shiloh, When I Was Young

IN THE SUMMER of 1972, at the age of 17, I was living in an old house on the corner of 20th Avenue and Ogden Street in Denver, close to downtown. Nobody used the term "homeless" back then, and I didn't have a sense of myself as someone who was particularly beggarly.

Nevertheless I was, and had been for a few weeks, without shelter. The downtown streets were populated with workers, out for a lunch stroll in the sparkling, transformative Denver sun, a sure rejuvenation from the closed-up offices where their workdays were spent. Clerks, businessmen, and poor working class, who performed janitorial work — their faces careworn as they waited for buses, joined in the fray.

Also, there were young people of all stripes. Some looked bedraggled, likely from another state. Others were suburban teens, giddily spare-changing along with the rest, or musicians with guitar cases open to accept donations for their music. Except for the occasional drunk person sleeping on the sidewalk in front of an abandoned building — buildings that have now been converted into hip, expensive shops and townhouses — you couldn't find the entrenched homelessness that exists today, a permanent fixture of many American cityscapes.

The young people that wandered the street in those days had of a mix of backgrounds. Some were escaping a rotten home life; some from comfortable, middle-class homes became disillusioned with their parents' (apparent) emphasis on capitalistic gain; others were merely taking a break from college, figuring things out. Most of them were ready prey for the many organizations that had sprung up to exploit young people in the hour of their pain and instability. Organizations such as the Moonies, the Hare Krishnas, and an abundance of quasi-religious outfits materialized to take advantage of the sense of emptiness in the lives of detached young people. Also, there were the everyday pimps, and a few honest organizations struggling to redeem the lives of disaffected youth, trying to gain them back before the lower elements of the street swallowed them whole.

The few friends and acquaintances that I had were "Jesus freaks"; the Jesus movement was sweeping the nation, and other parts of the world. This particular brand of young believer could be recognized by the small Bibles they carried in knitted pouches on their belts and their sober, or occasionally blissed-out, demeanor.

Always ready to quote Scripture for whatever question or situation might arise, they were a solemn contradiction to their free-living hippie contemporaries.

One of these friends, a teenage boy, knew of a place on east Colfax called the Paraclete; I believe that means Holy Spirit, particularly in His role as advocate, counselor, intercessor. Anyway, my Jesus freak friend and I wandered into the small bookstore to ask about shelter. The sweetly devout young man behind the counter looked up from his reading to suggest we try the Shiloh House. When we arrived at the Shiloh House, a tall, darkly bearded man questioned us, attempting to detect through a stern, brooding, scrutinizing stare, if we were sufficiently serious about the faith. He instructed us to go to our separate houses; one house was for the women, the one across the street for the men.

The women's house was run by his wife, Judy, a

women in her late thirties who wore long skirts and was smothered with an enormous mop of long, dark, coarse, messy hair, beyond which it was nearly impossible to see her face. It was my first experience stepping into an early twentieth century house; all of the houses I had previously resided in were modest frame houses.

This house had once been a large, grand dame of a house. At that point in my life, I was too young to see the features of romance and beauty in the sturdy wood floors; the wonderful auxiliary rooms, like a parlor, a mudroom, and a foyer; wood fireplaces; and the charm of the spacious galley kitchen with beadboard wainscoting. No, the house was dingy and cold, even in the summer, the furniture bare and ratty. And old. *Yuck.*

I was sent upstairs with my few belongings to find a bed in the dormitory-like rooms. Sometime in the next couple of days, one of the more expeditious members of the household took all of my clothing and either gave it away or repurposed it. The pretty, plum-colored velvet dress that I had worn to the Homecoming dance was cut up and made into a sewing bag. I was told that it was "too short" for a godly women to wear. There was a sense that I better shut up and go along with whatever was happening.

ALL MEMBERS OF the household were expected to work, and contribute all of their wages to the house; everyone was permitted to keep 10 percent of their earnings. Also, we had Bible studies and meetings a few times a week; these meetings were run by the big, bearded man who greeted us, the leader of the houses. Judy had several of her own children to look after.

I RECALL ONE young woman who stayed at the house; she had short, softly red hair, and liked to dab fragrant strawberry gloss on her lips as she looked in the little mirror over the sink. With my current adult awareness, I can see that she had likely suffered severe trauma, somewhere along the way; she spoke in a very small voice and displayed a wooden emotionlessness. She was about 19 and worked for a hospital as an orderly just down the road, maybe Pres St. Lukes.

We were friends, in that easy, just talking with each other kind of way that young people enjoy. She was sweet and friendly, and bore a quiet demeanor perfectly suited to the social climate of that house. She didn't seem to have any past, and her meek outlook appeared to encompass no more than her concerns regarding her daily shift at the hospital. She gladly turned over her wages to the house, happy to exercise her devotion to her faith in that offering.

I never quite understood what was going on, in an overall sense, in that house; I couldn't quite bring myself to trust or embrace the nebulous objectives of the house, or its managers and residents. So I did what I often do — I quietly observed. I feel protected in that mode. I become an anthropologist: a scientist gathering info, ferreting out facts, subterfuge that might reveal life-preserving truth behind a facade.

One evening, Judy asked me to deliver a note to her husband at the men's house. When I arrived at the house, I was told to take a seat and wait in the little entryway outside of his office. I could hear him talking on the phone in a boisterous and jolly tone that he never displayed around members of the house, his big body joyously moving around as he laughed and leaned back in his squeaky office chair.

When he concluded his call, he stepped outside of his office to discover me sitting there. He looked astonished, and then angry. I handed him the note that Judy wanted me to deliver and returned to the women's house. Shortly thereafter, he came roaring into the women's house. He demanded that I go out the backyard gate into the alley. There, I was met by my friend who had accompanied me to the house. The bearded man began shouting at both of us, waving his large arms in the air, gesturing and oppressively bending his massive frame over us, yelling that we were "marked from the ministry!" He didn't seem to care that neighbors could hear. Perhaps he wanted them to hear; perhaps he wanted to make a show of his authority.

Neither one of us understood what "marked" meant. We did understand that we had to leave the house, though we did not know why. Our belongings were brought out in small, black trash bags and handed to us. This man had such a steroidal physical presence that the other teen and I were pee-in-our-pants terrified of what appeared to be the ravings of a mad man, an ogre. We didn't know what he was capable of; could he beat us up, put us in jail? It was about dusk, the streetlights just beginning to cast shadows. My friend returned to his family in a Denver suburb. As for myself, whatever budding faith that I might have contemplated or hoped for was now completely quashed — thoroughly quieted to preserve my sense of what was knowable, dependable, and real.

Somehow I paired up with another young woman from the house. She was about 22 with dark olive skin and dark hair. She seemed sensible, and we decided to take what combined money we had, which was under $100, and try to find work and survive. We found an old apartment building to live in. It was sparsely furnished and included a bed and linens. The old radiator furnace made it nearly impossible to sleep, with its enormously loud, deep, otherworldly clanging and all night hissing.

Although I was unaware of it at the time, I am sure this place must have been shabby. And while I was concerned about my future, I was not yet panicked. Throughout my life, from my youngest years, my most pressing and guiding desire was to attend college. I am so glad I did not know at that time that I didn't have a chance in hell of making it to the steps of a college door from where I was; I was blissfully ignorant of how humble were my means for attaining much of anything that looked like security or hope.

At the grocery store, we agreed that a cheap loaf of bagged bread and a jar of combined peanut butter and jelly could tide us over for a few days. She told me that the reason we were sent away from the house was that I had overheard the manager talk about how he and his wife had gone with some friends to see the movie *MASH.* Apparently he feared that I would blow his cover because the people in the house were to assume a denial of the world and its distractions, including movies. I was completely unaware that the ministry enforced or expected any such restrictions, or that going to a movie proved any reportable wrongdoing on the part of the manager.

We set out to look for work at temporary agencies, mostly walking to wherever we needed to be. A couple of weeks into our venture, she decided to work at a bar on Colfax, a few miles away, just east of Colorado Boulevard. I got the sense that she might be in trouble, or mixed up with the management of the bar in some unfavorable way. She seemed gentle, and I worried about her well-being.

Our time at the apartment ran out, and we parted ways. I recall standing at an intersection somewhere along what is now Park Avenue. That particular intersection had five streets crossing. Clutching that small black bag of belongings, I stood in the bright, early morning Colorado sun, the season just starting to turn cold. I was cold, and so very, very alone. I was starting to experience a deep and overwhelming sense of anomie.

Trembling, I felt no connection to anyone or anything ... powerless. My sense of reality, built on the memories of the people that I so completely loved, seemed like a gauzy, temporary dream that lacked any substance or foundation — nothing left to sustain me, nothing solid enough to go forward with. Maybe, I thought, I could create a new direction. As an American, I had been trained to believe that I could wrest a future and hope out of the thin air of sheer will. But how ... the view before me was aimless and I was empty of knowledge or connection. Cars whizzed by me, filled with purposeful people, but what did I have to do with them, or they with me?

CAROL COVINGTON

Unpacking Daddy

This story is dedicated to my father, Technical Sergeant William Edward Covington, who served two tours in Vietnam in an effort to rise in rank while serving during a Hazardous Duty assignment. He ran from the cotton fields of North Carolina, enlisting at the age of 14 in the Army Air Corps by forging his birth record. He retired at Keesler Air Force Base, where he is buried after 25 years of service.

IN THE SUMMER of 1969, my mom, sisters, and I picked my dad up from the airport at Lindsey Air Station. He was returning from what would come to be his first tour at Tan Son Nhut Air Base in Vietnam.

Your body returned home long before the rest of you. Everything about you looked familiar except for your eyes. They looked cloudy, and I wasn't sure but

I figured you had some sleep in them. I remember thinking to myself how lucky I was that you made it home. Everyone in my class at Wiesbaden Junior High watched the news every night so that we could turn in our current event project due daily. Being military brats, we all shared the same story. Either our dad was leaving for Vietnam, already in Vietnam, or MIA in Vietnam. We knew whose father was never coming back, because they had left without saying goodbye to return stateside. Given the scenes I had seen on the news and the dreaded tone in the voices of Morley Safer and Walter Cronkite, I went to bed each night for more than a year terrified that we'd be the next family to pack up after the Red Cross station wagon pulled into the driveway.

That day at the airport, I looked at your hard-boiled egg–shaped face with the wiry gray whiskers. I didn't remember you having gray hair. I circled you in a precautionary way, allowing my eyes to scan the creases of your chin, elbow, and knees, looking for familiar scars like the one you got trying to add water to the radiator in the car while the engine was hot. I looked for evidence of wounds you may have gotten while you were at war. I listened closely on that first day when we all sat down to watch "I Love Lucy." I caught myself holding my breath trying to be as quiet as I could, just in case you laughed out loud like you used to. I didn't want to miss it, but you were more quiet than I was.

Nobody said that you might change. Mama had said for me to be good and that if I did what she told me and worked hard in school and didn't fight with my sisters, you'd come home. So I believed that you'd return the same way you were on the night before you went away. You had grabbed your duffel bag and the fried bologna sandwiches, which Mommy put in my Jetsons metal lunch box so that you could take something of mine with you. I put the last Scooter Pie in there too. I cleaned out my thermos really good so that the picture of George Jetson and his daughter Judy would help protect you, somehow. "Hey Dad, where is Vietnam anyways?" I asked. "Nowhere you ever wanna be," you said. I agreed.

Those months without you were hard. I didn't have anyone to play checkers with, or listen to the Yankees with, or cut the grass with. (I'm not supposed to tell you that mom's new friend cut our grass while you were away.) I watched every episode of "Wagon Train" after dinner while you were gone so that I could tell you if they ever made it to wherever they were heading.

After you returned, I pretended not to notice that something was different. I decided that it was just to "grown" for me to understand, and that sooner or later, things would go back to the way they were before. One night I woke up to the sound of crying. I know what Mommy and my sisters sound like when they cry, because they sound just like me. This sound was different. I stood in the corner of the kitchen between the bathroom and my bedroom. I saw your house slippers poking out from underneath the kitchen table, and I realized that the ghostlike moaning was coming from you. I had never heard you cry before. I tiptoed back to my room and slid into the corner of my closet with my blanket and tried to un-hear that sound. I hoped it was a bad dream brought on by that extra glass of Ovaltine I wasn't supposed to drink before bed. In the morning, I wanted to wake up and find you standing near the front door doing up those shiny buttons on your dress blues. You were not there, and it wasn't long before I knew that the war in Vietnam got off the plane when you did.

I waited for my dad to return home from Vietnam for more than 30 years. He finally came home on the day that he died, two weeks after his 71st birthday, when his soul and his body were at peace.

JAMES SPEED HENSINGER

That Damn Flag

THERE IT WAS. Every morning I drove past it. Every evening I drove past it going the other way. It was on a very tall pole in front of an apartment complex with a fountain and a duck pond nearby. Day after day, month after month, it was getting more and more ragged. It was in tatters. The red and white stripes were slowly disintegrating. It had started with the seam at the end breaking and streaming in the wind. Now the feathers of fabric were halfway back to the field of stars.

The owners were probably using it as an advertisement. I had read that the city banned commercial outdoor signage over 15 feet high, but flags were exempted because of the First Amendment. There were car dealers on Havana Street who competed with each other to see who could display the bigger flag.

One Saturday night after the bars had closed in

Glendale, I was driving home on Leetsdale Road and there it was. This time I decided to do something. I parked my truck and grabbed a pair of limb loppers from the toolbox. Holding them against my thigh, I walked to the base of the pole. The bottom of the halyard was caged in a padlocked box, but the loppers cut right through both lines just above it. I grabbed the lines, and pulled on the lower leg. After a foot or two, it started falling. I had to drop the loppers to catch the flag as it dropped. I had to catch it. She couldn't be allowed to touch the ground. Wadding it in handfuls against my chest, I worked the snap links off. Then picking up my loppers, I hurried back to the truck.

Three days later, when I was sober, I built a charcoal fire in my grill and burned it. No salute. No music. Just a few minutes of silence as I poked it with tongs to make sure it all burned. It was the best I could do.

American flag. Photograph by the author.

DAVID LENNON

Tulega

LOCATED ON THE Pacific Coast 70 miles south of L.A. lies massive Camp Pendleton, home of the 1st Marine Division. Since WWII, thousands of marines have passed through its beaches, canyons, and mountains. On a cold, rainy Thursday night in February 1960 a conclave of five recently arrived PFCs are sitting or standing around a pitiful oil-burner furnace — small and smelly — hoping for some heat. Small talk ranges from hometown stories — girls, football games, and family — to boot camp and ITR experiences. All hands had been meritoriously promoted to PFC out of boot camp headed for MOS position in administration, supply and logistics, and electronics and mechanical heavy equipment operations. The other marines were from Chicago, Detroit, Milwaukee, and Cajun country southern bayou Louisiana.

Conversation slowly moved on to Camp Pendleton's Camp Tulega Duty Station for the 1st Pioneer Battalion — a combat engineering outfit equipped for battle engineering utilizing bulldozers, road graders, and other assorted construction equipment, plus the capability for mining operations, anti-mine operations, and the role of EOD. The battalion had recent high-level success and renown from fighting in Korea, notably the battling escape of the entire 1st Marine Division from an entrapment by eight divisions of the Chinese Communist Army at the Chosen Reservoir, located close to the North Korean border with China and the USSR. The campaign had been fought in vicious far-below zero weather conditions.

This group of "boot" PFCs were assigned to headquarter company duties: administration, supply and logistics, and communication support. They had been "aboard" Camp Tulega for several months and exposed to the stories about Korea and also the Okinawa battle of WWII from salty old veterans. The camp was located at the very north most point of Camp Pendleton, about six miles north of Baseline Road, the major roadway through the base in that era. Located at that road that was the Infantry Training Regiment compound where all recruits from San Diego MCRD went for infantry training, the old WWII tent camp. Camp Tulega was isolated and off the beaten path of the division, served only by several civilian buses and USMC support vehicles daily.

The PFCs at this time are laughing and talking, bonding, aware that they would be serving together for several years. So far no acknowledged leader has evolved. PFC McCully was from Chicago, had spent two years at University of Illinois studying engineering. PFC Hunsinger was a Polish chap from Hamtramack, Michigan. Schmitz represented Milwaukee, Wisconsin. PFC Guidry, the only Jarhead out of the South from the group, was from New Iberia, Louisiana — a Cajun. He was also a 6' 4" heavy equipment driver. And PFC David Fitzgerald was a Colorado mountain lad from Old Snowmass in Pitkin County on the Western Slope — ranch and cowboy country.

The sole method of escape or entertainment at Tulega was the slop chute where beer, some food, a juke box, and pinball machines existed. There was a special area for the NCOs with a higher level with the same offerings. All the buildings at Camp Tulega were WWII vintage Quonset huts, metal semicircle tubes halved on concrete slabs. Cold in winter, hot in summer — perfect for the USMC's purposes. To keep the troops unsettled and uncomfortable.

In the short time that this group had been in camp, as they had all arrived at the same time in a USMC cattle car, a troop trailer hauled behind a semitruck, they had slowly merged into a fairly cohesive group. Going to the mess hall together, forming to formation for a.m. muster, and even occasionally going on liberty together. The remaining occupants of the Quonset hut were Sergeant Bly, who was a recently shipped over marine; Corporal Russell Corn, who had lost a stripe on Okinawa in some unknown violation who was very close to his discharge date; and Corporal Roberts, who was enrolled in Orange Coast JC and had a local California girl for a friend.

As usual, the five PFCs finally rack out at 2200 hours, as the bugle recording sounds taps over the camp loudspeaker system — grainy and scratchy from overuse. Sleep comes slowly in the chill air.

Suddenly, around 0215, the west door by the NCO area blasts open, lights snap on and an obviously drunken Corporal Corn stumbles in. As the lights snap on Corporal Corn bellows, "Wake up you boot sons of bitches!" All the PFCs are immediately awake. "Three more days of sand and grass, then Camp Pendleton can kiss my ass," Corporal Corn continues his drunken tirade. "Get up you worthless boots and show some respect to a departing NCO!"

"Be quiet and turn out the lights!" someone shouts back.

"All of you get the hell up!" Corn hollers.

Slowly we climb out of the cramped racks. Corn continues his ranting until PFC McCully yells, "Leave us alone. We just want our sleep!"

"Hey n—r, who asked for your opinion?" Corn slurs.

Instant bedlam and confusion erupts as all five PFCs take a confrontational stance, yelling at Corn that he's had too much to drink and needs to hit the rack. Corn yelling, "All you boots shut the fuck up, especially the n—r. I am so damn glad to be getting out of this miserable crotch, so just shut up."

The second explosion of the n-word enrages PFC McCully. He pulls his 14-inch bayonet out of the scabbard attached to his field transport pack, hung from his rack, and starts towards Corn. All the PFCs gather in a group, and Fitzgerald grabs McCully by the arm, "Don't do it, Walt. Don't let that chicken shitbird get you into serious trouble."

Wrestling, pushing, the young PFCs try to separate the angry two and keep McCully's bayonet out of the fight. McCully shakes and sobs, "Come into this goddamn Marine Corps and have to put up with this stupid, chickenshit racism from this dumbass Missouri redneck."

Sergeant Bly finally speaks up. "All right, let's all calm down, goddammit. Corn, shut up and get into your rack. We'll square this away in the morning."

McCully slowly puts his bayonet back into the scabbard on his pack, looks at Corn with a long bitter, withering stare. The Quonset hut is quiet. Everyone racks out. Bly turns off the lights.

The next morning, no one says a word. Nothing is ever said or mentioned of the incident again.

LUKE ANTHONY ALFONSO MARTÍNEZ

My Great Great Grandfather: Juan de Jesús Jaramillo

I dedicate this story to my great great grandparents. Their lives have touched me in so many ways. Their efforts to survive and thrive continue to amaze me and for that I owe them so much. This journey of exploration has taken me to places I never dreamed of going. I am blessed to be one of their descendants.

We don't inherit the earth from our ancestors.
We borrow it from our grandchildren.
Native American saying

New México Civil War Battles

NEW MÉXICO IS known as the Land of Enchantment. I am happy to tell this story of an enchanting family in an enchanting land. The realities of the Civil War and its many repercussions for New México citizens is reflected on the tombstones throughout the state. Several times, while visiting my ancestor's final resting place in Vallecitos, I had wondered what kind of a person he was. How did he endure so much adversity in his life? His personal character and patience to overcome obstacles say a lot about him and how he represented himself, his rural community, and his nation when it was being torn apart by abolition. It's not well-known how involved New México citizens were during the Civil War era.

The Civil War started on April 12, 1861. Leading up to the Civil War, many New México forts were established including Fort Fillmore, Fort Stanton, Fort Conrad, and Fort Craig. The posts safeguarded travel routes and created a deterrent to hostile Apache, Navajo, and Comanche raiders. On July 1, 1861, Lieutenant Colonel John Robert Baylor and his Confederate soldiers captured Fort Bliss. They used the base to launch attacks into New México and Arizona. Baylor and 300 soldiers arrived at Fort Fillmore the evening of July 24, 1862. Major Isaac Lynde commanded 492 Union troops. For the next two days, during the Battle of Mesilla, the Union side suffered injuries and lost four lives. They abandoned their fort at night and fled to the nearby

Organ Mountains through San Agustín Pass. This rugged mountain range has peaks that reach up to 9,000 feet. When Baylor discovered the fort had been abandoned, part of the Confederate force pursued Lynde's Army. Baylor accepted Lynde's surrender, winning the battle. Baylor proclaimed Arizona Territory part of the Confederate States of America and named himself governor.

Fort Conrad was mostly abandoned on March 31, 1854. It continued as a farming area for animal fodder, giving it the name of Hay Camp. Nine miles south of Hay Camp, along the Río Grande, Fort Craig was built as the military base for the 2nd Dragoons and the 3rd Infantry. During the early months of the Civil War, Fort Craig was protected by 3,800 Federal regulars, militia, and New México Volunteers, commanded by Colonel E.R.S. Canby. During the Battle of Valverde, Confederate forces attempted to bypass Fort Craig. The Union Army's objective was to prevent the Confederacy from crossing the river. The Union Army caught up with the opposition and fighting ensued. My great great grandfather Juan Jaramillo and his older brother Francisco fought bravely in the Civil War Battle of Valverde. The battle, in 1862, lasted two days. The Confederate forces, commanded by Colonel Henry Hopkins, suffered many casualties, as did the Union Army. Sixty-eight were killed in the Union Army. Thirty-five were never found and 160 were wounded. Within the Confederate ranks, 31 died, 154 were wounded, and one went missing from 2,600 troops. The Civil War continued until 1865. New México Volunteers of the Union Army were important in the defense of Southwest territory at the Battle of Valverde and the Battle of Glorieta. Both of these conflicts, as well as the Battle of Mesilla, took place in New México.

The Injury

AFTER THE BATTLE, some of the New México Volunteers accompanied a wagon train of commissary supplies heading north. The Jaramillo brothers were members of the First Regiment. They were in their early twenties, sons of María Beatriz Jaramillo. As the caravan made its way, one soldier, Seledón de Herrera, tripped on the trail, accidentally discharging his weapon. The bullet wounded Juan de Jesús Jaramillo in the right foot. Juan was transported to a military hospital where he spent several months in recovery. On May 31, 1862, he was honorably discharged as a private at

Polvadera, New México. He returned to Vallecitos in Río Arriba County where he farmed and tended sheep. During the winter months, in the mountains of northern New México Territory, Juan's wound worsened. Mobility became an issue as gangrene deteriorated the foot tissue. Juan visited doctor after doctor until he was awarded a meager pension. The stipend granted relief but life was far from easy.

According to a general affidavit in the pension records of Juan de Jesús Jaramillo, Nestor Martínez, age 56 and a resident of Vallecitos, New México personally appeared before a notary public to make the following sworn testimony:

> I was present as guard for a Commissary train near Polvadera, New México on or about the 12 day of April 1862 when a gun fell from the hands of a comrade who had stumbled and fell and was discharged, the ball going through the foot (right foot) of Comrade Juan Jaramillo the applicant in this case together with two of the three buckshot that the cartridge contained. I was present at the time of the accident and nursed Comrade Jaramillo about one month while he was in the

> Camp Hospital. The only thing I do not recollect exactly is the day of the month nor am I sure whether it was in the last days of March or first days of April but as well as I remember it was a month or perhaps a little more after we fought the Texans at Valverde.

Family Life in Vallecitos

JUAN MARRIED MARÍA Andrella Martínez at the parish church of San Juan Nepomuseno in El Rito on January 7, 1871. María Andrella was the daughter of Juan Manuel Martínez and María de la Concepción Lucero. The couple began having children, starting with Isabel, later in 1871, then José Eligio on July 29, 1872. As the size of the family increased, so did the military pension. George was born about 1873. María Cleofes came on December 15, 1875 and Bernardino was born April 14, 1877. María Rosa was delivered on November 7, 1879. The couple worked hard to raise their children and their farm animals while working the land in the mountainous southwest. Celestino arrived on May 30, 1885. He died as a young man during a torrential storm. He was traveling home along a dirt road next to the Vallecitos River. The flooding caused a bridge to wash out near La Madera, a village between Ojo Caliente and Vallecitos. The survival of the three youngest children, Fidel, Marina, and Juan Manuel was a true testament to the fortitude and endurance of the matriarch who gave birth through three decades of her life.

Like generations before them, the Jaramillo family adapted to fluctuating, seasonal temperatures in territorial New México. Unfortunately, living conditions in this time period led to a high rate of infant mortality. The loss of some of the couple's children included Delfino, María Concepción, Benigno, Juan Bautista, Malaquías, Cándido, Margarita, Amarante, and others. Common causes of infant or childhood death during the nineteenth century were breech birth, premature birth, miscarriage, inexperienced midwives, influenza, pneumonia, the common cold, sudden infant death syndrome, accidental poisoning, snake bites, spider bites, bee stings, tick bites, bear and mountain lion attacks, rabies, exposure to the elements, drowning, lightening strikes, house fires, and especially inadequate access to medical facilities. Even to this day, Vallecitos is located in a remote part of northern New México.

There were no clinics or hospitals in the area. As soon as a child was born, there was an immediacy to have the infant baptized. Sometimes the parents had to travel many miles to the nearest church in order to find a priest to perform the ritual. If the baby was born during the cold months, the offspring might be exposed to frigid temperatures en route to the baptism. Some mothers were not capable of caring for their children. Adoption among families was common. Not producing enough milk to keep a toddler alive led to malnourishment. The fact that seven children survived to adulthood is remarkable.

The Teller Institute

HAVING A LARGE family ensured plenty of helping hands on the ranch, but keeping the family together had its challenges. The older offspring began to marry and have children of their own. The youngest two were sent to the Teller Institute, a Native American School two miles east of Grand Junction, Colorado. The institute was on a 160-acre site and named for Henry Teller, interior secretary at the time.

The school opened its doors to male students in 1885, instructing 30, mostly Ute pupils. Pueblo Indians and other tribes came next. The concept behind the Indian school was a shared one, implemented in different parts of the country. These institutions attempted to give indigenous people, including the Jaramillo youth, new learning opportunities. Reading, writing, and arithmetic were the main curriculum but other topics of study included shoemaking, saddle and harness making, and agriculture. A few years later, when female students were accepted, courses in dressmaking and home economics were added.

On some levels the Indian schools were successful but on other levels there were problems, such as the loss of Native American identity, loss of family structure, and loss of cultural norms and indigenous language. The Teller Institute had multiple problems of its own. The plumbing in the school failed and the waste from the school became a controversial issue. The buildings fell into disrepair. After 26 years of education, the school closed its doors in 1911, the year before New México gained statehood.

Descendants

AFTER YEARS OF enduring a multitude of health problems, Juan had his right foot amputated on April 27, 1905. His foot was removed just above the right ankle and he was fitted with a wooden prosthesis. The pension again increased and Juan returned to a more sedentary life at home. Eligio and his wife, Juanita Pabla Chacón, helped care for the aging parents while raising their own large family. Juan died on September 11, 1910. The veterans' benefit transferred to his wife upon his death. Andrella succumbed to pneumonia on September 4, 1920. Their family, as well as the next generation of descendants, became a clan.

Isabel

The eldest daughter, Isabel married Antonio María Trujillo, son of Antonio Candelário Trujillo and María Soledad Valdez on April 2, 1888 in Vallecitos. They had the following children:

José Bernardino born May 20, 1888 and died September 10, 1919

Antonio Candelário, Junior born May 10, 1890, date of death unknown

María Elena born August 18, 1892 and died October 5, 1946

María Andrea born January 4, 1894 and died June 6, 1962

María Zenaida born December 15, 1897 and died October 1968
María Aurelia born November 12, 1898 and died September 14, 1920
María Teófila born February 3, 1903 and died January 14, 1986

Isabel and Antonio are buried in the front part of the Catholic Cemetery in Vallecitos. Juan de Jesús is buried in the older part of the same cemetery.

José Eligio

The eldest son, José Eligio, married María Juanita Pabla Chacón, daughter of José Román Chacón and Juliana Jáquez, on August 23, 1895 in Vallecitos. They had the following children:

Floripa born January 1896 and died in 1896 or 1897
Cirila born 1898 and died in 1899
Catarina born February 21, 1899 and died May of 1928
Juan José born April 27, 1901 and died January 19, 1980
José Ramos born April 27, 1903 and died June 5, 1958
Elacio Lafayette born April 24, 1905 and died January 1984

Simodosea born December 11, 1906 and died May 21, 1995

Alfonso Barcelón born November 9, 1909 and died January 12, 1970

José Perfecto born July 3, 1910 and died as an infant

Perfecto Saviano born March 12, 1912 and died April 8, 1958

Juan Ramón born March 12, 1912 and died May 19, 1959

María Vivianita born March 23, 1914 and died January 5, 2008

José Eligio, Junior born July 29, 1916 and died as an infant

José Isauro born March 23, 1919 and died December 6, 1989

María Anita born July 28, 1921 and died January 18, 1999

Salomón born July 29, 1921 and died March 15, 1931

María Graciola born January 15, 1926 and died September 4, 1926

Eligio and Juanita carried on the family business of raising livestock, including sheep, cattle, horses, and foul. They acquired property 5 miles northwest of

Vallecitos in a small village called Cañón de Vallecitos, now called Cañón or Cañón Plaza. In addition to the estate in Cañón, Eligio and Juanita acquired a 160-acre homestead at an elevation of 8,500 feet. The Jaramillo ranch, El Valle de Escondido, is nestled in the Kit Carson National Forest. Every spring Eligio and Juanita moved their household and ranch animals to El Valle for birthing. This summer home is surrounded by tall evergreen trees. With plenty of snowmelt, the property has two ponds. They pastured farm animals there through the summer, and then they herded them back to Cañón in the fall, a yearly ritual that continues today.

They sheared the sheep every year. They took the wool to Antonito, Colorado to barter for supplies of food, clothing, and other necessities. Winters in Cañón could be long. The Jaramillo Ranch still thrives through efforts of descendants fiercely guarding a way of life. Many moved away seeking employment in urban areas but the family cherishes its stories of a proud and patriotic lineage.

María Cleofes

María Cleofes married Fidel de Vargas, son of Querino de Vargas and María del Rosario Trujillo, on

November 27, 1897 in El Rito. They had the following children:

Querino born August 1, 1899 and died May 10, 1935
José Eligio born January 17, 1901, date of death unknown
José Dionísio born September 19, 1902 and died January 1987
María del Rosario born April 17, 1904, date of death unknown
María Margarita born March 29, 1906 and died in 1967
Antonio born April 14, 1909, date of death unknown
Rubén Cirilo, date of birth unknown, date of death unknown
Lugarda born 1909, date of death unknown
Victoriano, date of birth unknown, died in 1979

Fidel

Fidel married Leonor Martínez, daughter of Adelaido Martínez and María Hihinia Griego, on November 28, 1913 in Vallecitos. They had the following children:

Adelaido born October 9, 1914 and died July 1981
Andreita born February 7, 1921 and died 2002

Dennis born 1923, date of death unknown
Federico Polito born April 4, 1929 and died September 24, 2010
Orlando born April 4, 1929 and died in 1935
Olivia Grace born November 23, 1931 and died December 6, 2013
Amabelize born April 2, 1934 and died February of 2004
Fidelito born July 14, 1942 and died July 16, 1953
Esther born in Vallecitos and died as a toddler

María Marina

María Marina married Manuel Alberto Chávez II, son of Manuel Chávez I and María Juana Herrera, on November 18, 1920 in Vallecitos. They had the following children:

Maximiliano born August 18, 1921 and died December 30th, 1961
Benigna born October 1, 1922, date of death unknown
José Pedro born August 8, 1923, date of death unknown
Cristóbal born April 23, 1926 and died July 28, 1973
Ruby Magrete born November 8, 1927, date of death unknown

Manuel Alberto III born November 8, 1937, date of death unknown
Ida Viola born August 7, 1941, date of death unknown

Juan Manuel

Juan Manuel married Darita Medina, daughter of Edumeño Medina and Fermiliana Trujillo, on April 23, 1917 in Vallecitos. They had the following children:

Simmie born February 28, 1931 and died December 26, 2000
Antonio María, date of birth unknown, and died in May of 1989
Juan Manuel, Junior born December 23, 1934 and died June 17, 2009

A World War I Draft Registration indicates Juan Manuel's leg prevented him from serving.

He registered as an Indian citizen.

The Legacy

As a child, Cañón Plaza and other parts of New México were imprinted on me, having been born in Vaughn, Guadalupe County, New México. I was fortunate to have spent time with my grandparents in and around Cañón. It is one of our family's ancestral homes. My mother, Elva Viola Jaramillo was the oldest daughter of Alfonso Barcelón Jaramillo, a middle child of Eligio and Juanita. My great-grandparent's home was still standing throughout my childhood. It was a connection to my family's ancestors. My grandfather built his home next door to his parent's home. He grew up around many of his first cousins on his mother's side and father's side. He never served in the military but he did complete a draft registration during World War II.

One can find many examples of military service in the lives of those who descended from Juan and Andrella Jaramillo. The name Jaramillo was passed down through blood but the honor and distinction of serving one's country was passed down through the example set by the Civil War veteran, Juan de Jesús Jaramillo. Emulating his sacrifice is a complement to his life and what he represented. A way of life inherited from past generations speaks volumes of an agrarian era that still exists in the Southwest. Jaramillo stock preserves culture through language, customs, animal husbandry, food, history, and traditions. When early Spanish explorers came to the New

World in the fifteenth and sixteenth centuries, they mixed with Native Americans, giving birth to a culture that preserved Spanish heritage and Native ways. The Jaramillo legacy is a gift to country, state, and family.

KRISTINE OTERO

Out of Death: The Birth of a Combat Veteran

WAR. IT DID so much so fast to who I was. It cannot be unseen or undone. As a result, relationships and the effort I put in to them are black and white, simple, but complicated in a way that very few understand. My ability to function as one part of any relationship truly comes down to a feeling of safety, loyalty, and a willingness to take a bullet for the other person. In my PTSD brain, this makes sense. To my family, it seems dramatic and unreasonable.

Prior to joining the army, I existed as an extroverted, irresponsible girl of 22. I will refer to her as "she" because we don't have much in common, except our shell. I didn't like her when she existed. I'm grateful she is gone.

It never even occurred to her that the country was

engaged in a war. She was desperate, grasping at anything that would hold her long enough to trap her and save her from her life. A contractual obligation to the army did just that. And more.

In training I felt her slipping away — changing — and not only the 24 pounds of muscle our shared shell gained. The world was concurrently dimming and expanding, autonomy and individuality disappearing, falling in line as a soldier. I was bad at most everything, and really clumsy. To be able to hide in the middle of a group only made me stronger. Honing my skill set, I was mouthy, fearless, and above all else, determined. In the army, you don't have to be the best, but it's a good idea not to be the worst. I made sure I could hold my own with the boys. I became one of the boys.

Misogyny, sexism, gender inequality — it all exists in the military, but I tried not to let it affect me. For females, for me, this meant I had to try harder at everything. I had to prove I was big enough, strong enough, and tough enough mentally. I didn't get help carrying or mounting my crew-served weapon. It was sink, swim, or get the fuck back and let the boys do it. My NCOs joked about selling me to the local nationals for a goat and two tomatoes. That was my worth; I only wanted one of the tomatoes. Prior to opening combat-arms jobs to the females who could pass the rigorous tests, we were thrust into them out of necessity. It wasn't newsworthy, it was combat. I saw the obvious in that if the uniforms were a one-cut-fits-all — so was the attitude. Having no desire to beat them, I preferred to join them.

By the time we deployed to Iraq, my world was so small all I could see were the front and rear gun trucks — of the convoy. Nothing else mattered.

A trusting, loving codependence exists in war. I was entwined in these relationships to the point of survival. By the end of my first 13-month deployment, I began to get separation anxiety from my battle buddy. There was a very real possibility we could be separated back in Germany. Soldiers were already receiving orders for new duty stations before we had even hit safe ground. My battle buddy, Joey Otero and I decided we loved each other enough to get married so we would at least be together for the next few years.

In many ways it was a marriage of convenience, which is common in the military. But it was rooted in love and our dependence on one another. I was afraid to be without him. He had my six, and I had his. We had a very real discussion, admitting neither of us knew what marriage was supposed to look like. Knowing I was a lesbian, but I would do my best, we agreed our marriage would be a partnership of two best friends willing to try their hardest for one another.

Only after the fact can I look back and emphatically say my husband saved my life. Having each other made the transition into the civilian world less shocking. We felt alone, together. Although I don't know for sure where the problems begin for veterans, I do know that very little of our thinking translates over to the mundane life after combat. Almost immediately I realized I operated in fight-or-flight mode all the time. I couldn't handle simple tasks; they made me nervous and unsure. Leaving the safety of my home was overwhelming. I began to plan my life in advance. I made daily lists for everything. Having fun, making plans, living life — ceased. It was too dangerous. There were too many variables to make leaving worth it. Crises, big or small, were the only situations that allowed my screaming brain to slow down, focus, and execute. I began to catastrophize everything as a means to function as normally as possible.

At the end of the day, Joey was by my side. I could verbalize exactly how I felt, relate it to a shared experience, and know he understood both my words and how it impacted me — without judgment. He had this in me as well. First and foremost, we were a team. My wedding ring and improved last name tethered me to the world, and to him. I needed that safety and security.

My damaged thought process made trust nearly impossible. I thought it was everyone else who was unwavering and unreasonable. My husband never faulted my logic or rigidness. He loved me through every tough day, and never argued with me. He justified my rapidly developing coping skills, never thinking I was broken — until I snapped.

We were at annual training in the Texas National Guard, where they had set up a mock FOB. Arriving there, I was instantly triggered and uncomfortable. It was very realistic, with river rocks and conex boxes. My brain couldn't shut off the hypervigilance that exists in Iraq. My adrenal glands were pumping nonstop. It didn't help that my M16 had a blank firing adapter on it. I needed ammunition to feel safe. One day I was napping and the Air Force began doing drills, and dropping bombs. It startled me awake on my cot as the conex rattled. I couldn't calm down; it felt and sounded like mortars. I thought I was having a heart attack.

I found Joey who felt my heart slamming in my chest and took me to the medic station. They wouldn't let me in the door. I became nauseous and took my Kevlar off thinking I was going to vomit into it. They took my blood pressure on the porch of the conex, and they wrote me a profile, for rest. They did nothing for me.

When we returned home, it was much of the same. My body was stuck in survival mode. We decided to go to the VA hospital. If you want to be taken seriously by the VA, walk into the emergency room in fight-or-flight mode. If you aren't ready to be bombarded and surrounded, try to calm yourself down first.

My brain was foggy, unable to focus, making me angry and weepy. My pounding heart drowned out much of the background noise, and my hypervigilance made it hard to stay still as my eyes darted from one thing to the next. I had tunnel vision. Everything was so hard. I realized after that I had been sequestered in a small room with a wall of windows and I couldn't remember the last time I had left the house. This was who I had become without being present for the transition.

The only calm in my world was to look up and see the face of Joey. I was scared, but I knew he would keep me safe — always. He would take a bullet for me.

He could read my body language, my facial expressions. He never pushed me harder than he knew I could bend. He knew my limitations in the civilian world better than I did. We never talked about it, it was simply who I had become — what I could and couldn't handle. Even as I was devolving into this person, I knew I was extremely lucky to have my solar system in one person.

My coping skills developed, and none of them were terribly detrimental. I operate in crisis mode almost all the time, but to the outside world it comes across as neurotic, a little obsessive-compulsive, and very bitchy. In my PTSD brain, if everything is life or death, there is no room for error, no in-between. Days must be regimented and follow a routine, a predetermined plan, and executed as if my survival depends on it.

It's extreme, it's taxing on my sympathetic nervous system and general health, but it keeps me calm — as calm as I can be. At this point, it's my normal. And as far as I can see ahead, it will still be my normal.

Different Isn't Better

FROM THE START of my journey through the army, I knew I was different from my peers. I didn't share the popular mind-set on anything. I had no love of country.

No familial obligation. No frame of reference for anything camouflaged. The American flag meant nothing to me.

We were going to fight in a war I didn't support. I didn't support George W. I was young and liberal, but without any fundamental convictions. I could have objected, but deploying had absolutely nothing to do with fighting a war. It had everything to do with protecting my brothers and sisters. I had melted into a platoon, and we breathed, moved, thought as a unit. Our leaders had done a good job of ensuring our dependence on one another.

I made the decision to bat for one team — the girls team. I didn't even know overt homosexuals were not allowed in the army. "Don't Ask Don't Tell" was still in place, and it never came up during the enlistment process. The first clue I had was watching a video on the illegality of sodomy while in Basic Training. It was a bizarre way for the military to denounce homosexuality, but eventually I understood the intent. I knew I couldn't walk around yelling "I'm a homo," although people could walk around yelling "I'm straight," but I stuck to my convictions — until I didn't.

I learned quickly that the military is a boys club. Play

by the rules, because to cry that everything was unfair would get you nowhere but noticed. We were all misogynists together. Not all females in the military wear real bras and makeup; some fall in line and call bullshit when necessary. We were all provided the same training, and the same amount of time, to be stripped of our autonomy and gender. By the time we get anywhere, we should all be the same.

I was different in other ways, too. Right before I deployed the first time, AFN (Armed Forces Network) was airing the capture of Saddam Hussein. I was sitting in my room in the barracks, alone. My door was open; the doors were always open. My peers were gathered in groups, watching it like a sporting event, cheering up and down the hallway. I sat on my bed clutching my knees to my chest, unable to tear my eyes away. My reasons were not the same. I hated what I was watching. I hated myself for watching it. I hated everyone else for thinking the dehumanization of this person was alright. It was like watching a snuff film.

The images are burned into my brain. He was dirty, his eyes were wide in terror, and he looked resigned. He was being examined on live television, and I couldn't understand why. My roommate ran in, and I realized

I had tears streaming down my face, with a scream caught in my throat. She panicked, not understanding why I was upset. I told her I thought it was disgusting, made worse by the cheers. She heard me, she understood, but she also supported the coverage. She was a proper soldier.

I realized our country is made up of voyeurs, desperate for justified retaliation. Saddam had wronged our nation — Papa Bush. He was the head of the snake, and under Junior's command we had caught him. To everyone else, that was all that mattered. It made our upcoming deployment so much more important to them. It made the thought of war a little less scary, a little more palatable.

I saw a scapegoat. An old man, broken, and scared. They said he had been found in a "spider hole," and I imagined myself in his position. I hurt for the predicament this man was in. I would never have verbalized my feelings inside the military. My thinking was wrong, I know that. We were justified in our capture, but the live footage was taking it too far. The USA has the biggest dick, and will take any chance to show it. It's gross. I knew it then, and I will say it over and over now, on paper.

I have never wanted to be lumped in with the generalization of how our nation imagines veterans. I'm not the veteran you can spot in a crowd, unless you have a keen eye for military body language. I hide my years of soldiering well — from most everyone. I don't tell people I'm a combat veteran. I'm a tiny tattooed liberal lesbian, and that's all people see. I like it that way.

I didn't do what you think I did. I didn't protect this country, or America's freedom. The war in Iraq literally had nothing to do with our sovereignty. That notion exists to show our veterans that the mistakes made after Vietnam will not be repeated. We are not owed a pedestal, though. I appreciate not being called a murderer, or spit on, but honestly, it would be justified. We occupied a country to rape their land, pillage their oil, and fulfill a grudge.

The Destruction of My War

War affects our country so deeply. It does not extend to simply the act of fighting and forcing generations to live with the consequences of wars, creating a status quo. It does not simply touch the soldiers fighting these wars as they ebb and flow, change tactics, stop getting reported on, or even end.

Knife wounds on top of existing scars. The soldiers fighting these wars are forever altered. Many families are torn apart as parents and spouses realize their soldier — their veteran — is different.

I am not an exception to this destruction. My relationships are damaged, some beyond repair. Most of the damage has been caused by my altered and broken way of thinking. People are only people, after all. Therefore, I have lost the ability to have unconditional love or obligatory relationships solely based on genetics. My family saw these changes. My family saw me, but I was not me. They treated me like I was me, but I was gone — different, mean, rigid, unwavering. It redefined the dynamics; it wrecked or rewired many personality traits my family was used to encountering with me. I still catch glances in my direction, as if my parents are looking for me inside of the shell of who I am now, outwardly. They don't want this version, they don't like her. But, she is me, and I am her, and I will never change back.

Desperation in Writing

A WELL-TOLD STORY opens the wounds of the narrator. Nothing worth saying is easy. I work hard to cut down into scars, bleeding my vulnerability onto the page. There is no room to regret an omission after the fact. If the end result makes my heart hurt, if I feel disgusted and relieved, I believe I have told my truth. War is not pretty, ever. War can be funny, the smell of dark humor lingers over a combat zone as strongly as fear and dead goat carcasses. The aftermath of a well-timed mortar attack in the field fills the air with laughter, the smell of soil, and something burning. If I close my eyes, I can hear it — smell it. I could be anywhere with a group of guys yelling homoerotic bullshit to one another about dog piling. Those memories are comforting; they make me smile when taken as the whole.

To tell my story felt very self-serving and egotistical, much like this piece feels as I write it. While earning my undergraduate degree, I realized how much I loved writing. I also discovered I was able to emote more successfully on paper. For a long time I couldn't justify a reason why I had a story worth reading, though. Who was I in the grand scheme of things? I was a lowly female soldier who never earned any rank past specialist, married her battle buddy to prevent separation, and ended up with severe anxiety-driven PTSD.

Once I began to write I realized it was an extremely

cathartic way to sift through the emotional impact of my war. Not only Operation Iraqi Freedom and Operation We Are Still Fucking Here, or whatever they called it as it progressed, but also the war I was waging with myself. My adrenaline would surge with each memory, as if they were reoccurring. My memories are triggers. They are muddy, loud. They have smells and feelings, and flash like frantic still shots from movies. I never try to stop them from coming because in the chaos and panic there is also safety, comfort, and predictability.

Writing about my experiences is the only way to slow my memories down, to make sense of my war, to quiet my mind. I write the bad days in a linear fashion, frame by frame, evoking the feelings and the smells. Writing allows me a second chance to go back and see the place, the days, the moments that fucking destroyed my humanity, stripped me down to my core, and created the version of me who exists today.

I couldn't see this when it actually happened. War is fast, there is no time for independent thought or feeling. Soldiers execute from conditioned response: survival. I never looked back at the end of a day in regret, only gratitude that none of my brothers and sisters were killed. I slept soundly in combat, sometimes not even the mortars landing 100 yards away would wake me.

Soldiers go to war to die, and most of them do — psychologically. I did. I only realized this once I began writing. There will always exist in me a small sense of loss for the girl who deployed. She was untainted. She was a lot of things I am not. Her innocence and ignorance deserved a proper burial, and I obliged. My humanity and compassion returns with every story I tell. I was given the opportunity to grieve my own death through my words. I am no less of a monster, a war criminal, but I am holding myself accountable by telling my truth.

Poems

MARYANNA WIENBROEER BRUNKHORST

Metal

A perfect harmony
of wind-fed heat and forge
to usefulness
shock and battery create
not life but potential

trapped inside the
bearer decides whose
she'll be
carrying as trained

Neutrally fine-tuned
to be deadly always cutting well-
honed for parting honied
sinews departing humanity
based on ephemera

Empty
metal
drains red
          tarnished.

Dog Tags

Fly with 'em, don't
wear 'em,
they don't burn but
may warp.

Airplane wing shot from a window looking into dawn sun. Photograph © James Speed Hensinger.

Honor

21 casings — bronze, oblong
deliverers of honor
and death. How
ironic — empty shells full
of memory — yet vacant.

Uniformity — reduced to
data, embossed metal —
all that's left. So
cold and barren, stripped
of personality, storied
humanity. Just the facts

"Ma'am, there's been ... "
Accident that the uniforms
come to break
the spirit's veneer of self-
respect, reality crashing
in. A flag, folded.

Honor

Next

I am waiting for
the rain.

Not the deluge that streaks the panes and darkens
skies, freezing
joy, drowning possibility, and deadening
the way. No
I am not awaiting that rain.

Nor am I waiting for the misty, barely there, secretive
blanket
of moisture's breath that hides from itself, unsure of its
right
to exist. No
I am not awaiting that rain either.

The rain for which I am waiting drips light, grows
greening, promises

sunshine freshening and quenching just enough to start
    a seed, caress
a dream. Yes
    That is the rain I'm awaiting.

Eraser

How innocent the tool you chose
to wield as weapon. A child's tool —
soft, flexible, easily overlooked.

I wonder though
What would you have told your
daughter if she had been in the room?

> "Let me borrow your eraser Honey, I need to
> fix something wrong with her score — it's too
> high."

> "But why Daddy? She did good, didn't she?
> So why are you taking it away?"

Was it because I'm desirable, but not
open for business? Or were you threatened
by my strength, my intelligence? My presence?

"There's a guy who we wanted
to be the best, Honey."

"But was he, Daddy? Better than her?"

"Well, no. Not really."

"So why then?"

Would you answer that it wasn't my place?
That flying is a man's world and
I needed to get back where I belonged?

"But Daddy, I thought you said I could be
anything I wanted?"

"You can Honey, you can."

"So if I can Daddy, then
why can't she?"

You erased me —
will you erase #her too?

ALICIA R. CHRISTOPHER

I Am Sorry That I Have Changed: A Love Letter

Dear my love,
I am sorry that I have changed.
I wish I could talk to you,
I wish things were the same.
I just feel like you will never understand.

As a soldier, I developed a strong sense of pride.
I tell myself, “You cannot be selfish, so keep your
emotions inside.”
But the truth, it shows up in ways that seem so unfair
Taking over my life in the screams, paranoia, and
nightmares.

So I assume and I wonder if you are sick of me.
Are you tired yet?

This life has become a roller-coaster ride, for us all.
I am sure you have your regrets.
We are supposed to be soul mates.
I am sorry to shut you out.
Trust is very difficult for me, now,
I have lost my way somehow.
You ask questions from time to time.
There are things you want to know,
I have had to leave parts of me behind,
Parts of my soul.
Even the better parts I once shared with you.
I left them over there, I thought they would be impossible to live up to.

Dear my love,
I am sorry that I have changed.
I wish I could talk to you,
I wish things were the same.
I just feel like you will never understand.

Remember the other night, you wanted us to hang out with friends?
I started a fight, I didn't want to go, and I didn't want to pretend.

You see, I live in my own world.
I fight the pain and the memories in my head.
It takes everything for me face another day,
It is hard to get out of the bed.

Dear my love,
I am sorry that I have changed.
I wish I could talk to you,
I wish things were the same.
I just feel like you will never understand.

Have you ever seen a lifeless body, how about two or
twelve?
Have you ever lost a friend or soldier you were
responsible for?
And were the only one left to tell?
I even questioned God, on why I survived.
Some of them were parents, I did not have kids,
So, why did I stay alive?

See, what I just said, it shook your core.
It is hard to share my memories,
I am afraid to open that door.

Sounds, smells, and anxieties come rushing back to my
mind.
It forces me to live it all again
Filling my subconscious mind, every time.

Dear my love,
I am sorry that I have changed.
I wish I could talk to you,
I wish things were the same.
I just feel like you will never understand.

Even in my joy, I don't crack a smile.
To you I must seem heartless, but if you give me a
chance
Maybe in a little while
Our life will be average again.

Dear my love, there is a void in me,
Maybe you have one too.
Please don't give up on me, it will take me some time
to share with you.
It is easier to walk through this life with someone
If you understand all that they have been through.
I promise to try, I promise to cry, I promise to release a

little at a time.
I just need your hand, please stay, and walk with me
Until I am able to be who I once was for you again.

JASON DAVIES

[In sacred space]

In sacred space,
I hold your world.

You explore
soul's temple
for treasure
and answers.

Hour ends.

Contained,
a tenement of souls,
waiting,
to be visited again.

[The soft glow of street lamps]

The soft glow of street lamps
illuminates my thoughts and dreams.

Neon signs advertise my desires
in whispered soft hum.

My fear walks the street
turning me out, a cheap trick.

I grasp to hold out
for gentrification.

[Trauma/ on a pulpit]

Trauma
on a pulpit
of pain
decreeing
salvation is gone.

A medicine man
peddling tonics
for mental ills
upon a soap box.
He barks
salvation is here
for a limited time

If you have the coin
for it.

ALEXANDRA JACKSON

Ecology is the study of the relationship between living organisms and their environment. An ecosystem is a place, such as a rotting log, a forest, or even a schoolyard, where interactions between living and non-living things occur. All living organisms and non-living parts within this place are interacting all the time and adjustments must occur if the organisms are to survive.

Geosystems: An Introduction to Physical Geography

Solider, disclose about your living relationship with the non-living. Solider, speak! How you wrap your fingers around the neck of the smooth non-living glass. How you wrap your sewed, shut, cracked, bleeding, living lips around the non-living lips, take in the elixir of life. Solider, tell me, how you use the non-living liquid to make you live again. How the pictures of the non-living dance around in your mind eternally leaving a dead scar across your living brain. How much of this chemical charge determines your survival? Solider, sing to me, the song of your dead soul.

Libera me, Domine,	Deliver me, O Lord
De morte aeterna	From eternal death
in die illa tremenda	On that awful day
quando coeli movend sunt et terra,	When heavens and earth shall be shaken
dum verneris judicare saeculum per ignem.	And you shall come to judge the world by fire.

Soilder, ***my*** brother, you are a decomposer of this world living on the

Dead

Organic Debris Left Behind

By a once living process.

[In Real Life]

I had a dream one night.
I came home, to our childhood house.
Envisioned as clear as it is now.
Clutter lining walls, cat fur suspended in air.
Up the stairs Mom weeping on her bed.
I walked into the cracked tile kitchen.
I glanced down at the copious amounts of carbonated
liquid sugar.
Wondering if I should ...
I heard your fast-heavy footsteps coming down
from your room on the side of the kitchen.
Like a bowling ball bouncing off the wall.
Children racing on tar black pavement.
Grin abeam kissing their faces
Crying out in laughter
You're out.
You swung open the door
like an oven being dropped open

Ungreased apparatus screaming as
Sizzles of crackled pig's ass skin
flood the air
Holiday feast.
You were dressed head to toe
in camouflage green,
your uniform,
Army strong.
You even had the soft-square hat tailored to your head.
I glanced down and, in your holdings,
A dark black shape, one that made my skin flush with
Cool nausea sweat pearls.
Your voice was screaming out — forming only sound
    waves,
rich high-pitched diaphragm screams.
Eyes rolling in and out of the back of your head.
You were marching, knees shooting in and out of the
    bottom of my
Tunnel vision.
I ran to the living room and screamed up the stairs to
    Mom.
HE'S GOT A GUN,   He's Got a Gun,   he's got a gun.
My body pressed against the corners of the cat litter
    sprinkled carpet steps.

Mom called down *What?*
Like an omen you were standing next to me
I turned my cheeks looking to you.
Along my gaze taking in the neon green cat toy bin.
Maybe if Mom hadn't cleaned the litter boxes when
you were developing.
Taking in your dessert storm canvas boots.
You tied them in perfect bows.
Is that how you wanted to be remembered when you
died?
Maybe if your trauma wasn't weaponized and locked
away.
Mom's grandfather clock seemed to bulge off the wall,
Hands twisted and pulsating.
Coming into focus
The handgun in your hand as it touches your forehead.
My stomach dropped out of me and
I buried my face in the wall of the hallway.
Every cell in my body tenses ready to take the damage.
I didn't want to awaken.

I saw the outside of the house
with the windows looking in.
Curtains drawn,

a second of a burst of light,
opens a portal to the skies.

propellant gases exiting the firearm behind the bullet —
Although amongst the brightest,
the heat of the flash dissipates quickly and

thus is no longer visible.

Then I awoke

Yet you were already dead.

MICHAEL MCANDREW

Prick

A needle prick in the arm
Junk up the nose
It shouldn't have been like this

She thinks back to how good she looked
In the uniform

Looking in the mirror
Even if her hair was in that stupid bun
What looks the best was how she saw herself

These fucking Whole Foods cupcakes
She knew she was gaining weight but didn't care
The junk made her crave cold sugar

She may as well pick up, try to score
It's not like she could drink after the DUIs
Intoxalock doesn't test for heroin

Maybe getting kicked out wasn't the worst thing
At least she didn't have anyone
Telling her what to do

Her boss

Her boss

Her boss

It didn't matter

The NJP had been quick
Nothing happened to him, but the whole command
knew
The women under him

Push it away, pick up, go home, shoot up
She had a dog and some disability fun bucks
Seems strange to be disabled from something

That never happened

Don't say it, keep quiet

The dog gave her cold wet kisses
Her nose ran too
She held him close and breathed in

Dave

Never ask Dave about how his wife is doing
He probably got divorced again
It just makes things in the smoke pit awkward

Never drink and drive
You'll absolutely make it past the cops
The gate guards on base are another story

Never date someone in your shop
Things will go really well for awhile
Until you fuck it up like you did with the last one

Never get married for BAH
It's a lot of extra money
For the divorce lawyer you'll have to pay for later

Never go on deployment
Eventually you'll have to come back

That divorce won't pay for itself

Never buy a brand new Mustang
18 percent APR is a really good deal though
The dealer even gave you "military financing"

Never get that moto tattoo
Is that an Eagle, Globe, and Anchor?
You're a cook in the Air Force for chrissakes

Never cheat on your wife (again)
Especially not with Dave's wife
See, this is the kind of shit that makes it awkward in
    the smoke pit

Helicopter outside of a village near Phan Thiet, Vietnam. Photograph © James Speed Hensinger.

Midrats, or, Fourth Meal

Hot biscuits and gravy
More or less snot on bread
Bacon
With the texture & consistency
Of an old scab smothered in ketchup
The eggs have the faint aroma
Of degreaser and disinfectant

A chicken patty
I get two from the guy
I cranked in the chief's mess with him before he got
kicked out
Luke warm already, one bun only
Skip the cheese, been sitting out
Cold seasoned fries
"What's the seasoning?"
Guy holds up plastic container marked "seasoning"

Fuck it
Don't press my luck with dinner
The line's too long anyway
Smoke a Newport
Chug coffee, black
Cram a Kit Kat in mouth, swallow some of the wrap-
per (just some)
A cheap buzz as I head up on deck
For the last watch of the night

Later on, I trade some pre-workout
NEW! LEMON LIME "FLAVOR"
To a CS2, "We good."
For a shrimp po'boy from the ward room
Might actually be fresh(?) shrimp(??)
The roll is soft, not even close to stale

This is the best meal all day
I think
As I wipe the horseradish away
From the corner of my mouth
I smile
Skip mid rats

Untitled

Moon
Coin on a dead eye
Only other light
The soft glow of the HUD

Hornet lazily drifts
Under lunar token

"On the ball"
Right between crosshairs
"Is he gonna catch?"

Fuck

Another pass
Wire gets skipped
Pilots lean back against the windscreen with me

Droning on

They could have made it, easy, one pass, easy

"On the ball"
Take two
Hits the deck hard

Caught it
Green shirts retract it
He begins to taxi

"Hey thanks a lot man!"
"You too sir, you too ... "
They all shake my hand, head down

Break down HUD
Time for a last cigarette
Then hit my rack

As I do
I glance over
And wonder

Where did those bombs go today?
I spit into the ocean
Not my problem

I Still Do

I watched a man
Drop his wedding ring
Into the ocean
On the eve of my first deployment

"You dropped it in the drink"
He said it was an accident
But
He divorced that year

That's not the reason I don't wear mine
I got married between deployments
And had a hard time believing
Anyone was waiting for me

Untitled 2

The inside of my mouth
Already tastes
Like beer vodka cigarettes vomit

Very calmly
I add the umami
Of a shotgun barrel to my palate

In this moment
I feel so Zen
Tears dry stomach quieted

My blood
Is going crazy
Staccato pounding in my ears

I'm outside on my porch
My wife sleeps inside

I don't want to make a mess

This is just a dry run
I didn't put the shells in yet
Not very brave, I guess

I put the gun away
Make a mental note
To try harder next week

Motherish

It occurs to me
Not for the first time
I climb into my coffin rack
More like the womb
Than the tomb
Strange kind of mother
Named after a dead president
One of a kind one of a dozen
A kind mother kind of mother
1092 feet long bound back to the Persian Gulf
The cradle of civilization
Her children fly from her
Full of fury
Just as quickly return
Slamming into her
With a frequency
That can only be described
As a repetitious toddler

Playing hide and seek
Ships are always women
Not always to suggest
A lover
The place a young man can call home
Bound tight in her steel embrace

Fiction

CRISOSTO APACHE

Headlights

IT WAS DARK out, and I was lying in my bed kicking back. The house was empty, except for my little sister watching television in the other room. My parents were out at some stupid dinner function sucking up to people with money. They were always trying to get ahead, and that is how we ended up in this stupid town. There was nothing to do here except drive around. My parents don't trust me driving around by myself, yet. They think I might jack someone up or never come home. If I was going to do that, I would have already done it. The air was still warm, my window was open, and I could feel the slight breeze settle, cooling the skimming beads on my forehead. The marauding song of the crickets against the night was interrupted by Jack pulling up in his piece of shit dust buster. Well, his brother's dust buster.

His brother was doing another stint in Desert Storm

and left him his truck to drive. Santa Fe was not a big city by any means, but much of the city was spread out. Jack was my way of getting around. His brother's truck was this beat-up yellow 1974 Ford pickup with rusty panels that rattled when we took the dirt roads. It was cool to ride in because he had these tires that made it ride high and roared against the pavement.

I jumped up and ran outside. I opened the door and hopped inside. Jack sat there with a cigarette hanging from his mouth. I never told him the cigarettes-hanging-from-his-mouth "look" made him look cheap, but I suppose I was not far from being wrong. He back-handed me on the shoulder as usual and we sped off into the darkness. We cruised the back roads of town and Jack backhanded me again. He told me to reach under the seat. My face became twisted and bewildered. My hands patted the bottom of the floor. There was a soft mass which felt elegant. Now I had a smile on my face because I thought Jack had stiffed another pair of his stepmother's underwear. I clasped the soft lump; my smile diminished. Against the softness, the lump quickly became rigid. I pulled up the wad of cloth and saw it was dirty and stained with grey and black streaks. I unfolded the layers; it was a pistol. The weighted grip was cold in one hand as I caressed the piece with the other, not knowing what one really felt like, to hold. My blood rushed into my hands, warming the butt into a comfortable persuasion. Jack let out a wary chuckle and looked at me with a half-smile.

"Where did you get this from?"

"Uhhh, I took it from my dad's shirt drawer?"

"What are you going to do with it?"

"I thought we could go try it out near the dump. Wouldn't that be sweet?"

I hesitated. I poured the gun back into its cloth and slowly tucked it away where I found it. As we drove down St. Francis Drive, the streetlights became a cantaloupe-colored blur. My stomach began to feel like I had gravel for dinner. We turned on to Cerrillos Road and headed east.

"Where are we going?"

"I need some smokes."

We pulled on to Guadalupe Street, and I noticed three dudes standing on the side of the Allsups convenience store. They appeared as long shadows against the light on the side of the building. Jack stepped on the gas, and the tires echoed a deep squeal.

"GOD DAMN, FUCKING FAGGOTS! You see

them everywhere these days!"

Jack pounded the steering wheel with his palms and began to breathe deeply. We pulled to the back of the store and came to an abrupt stop. Jack opened the door, leaving the truck running, and made an irate dash to the guy wandering behind the store. Jack pushed him, grabbed him by the neck, and threw him to the ground and kicked him. I noticed one of the larger guys heading towards Jack. I jumped out of the truck, ran towards him, and pushed him from behind. The gravel dinner in my stomach shifted. I knew I was going to get my ass kicked. This guy was much larger than I was, and he did not seem like a "faggot" to me. I took a step back and made a run for the truck. I reached underneath the seat and pulled out the gun. I pointed that sucker right at his face. If he had known I was scared to pull the trigger, I would be a bloody pulp.

"Common, faggot! Do something! COME ON, DO SOMETHING!"

He turned and told his friend to go inside to call the police. I could only see Jack as a shadow behind the store. All I heard was the thudding of his kicks. I saw Jack come into the light covered in blood.

"GET IN THE FUCKING TRUCK!"

I turned and jumped back into the truck and slammed the door.

"What the FUCK was that?"

"Just doing my part to save our planet from fucking faggots!"

My head lowered. I began to sink. Drops of sweat fell from my forehead. I looked into the rearview mirror and saw a crowd of people pooling outside. I saw the big guy standing, watching us speed away. *Ahhh, the rocks!*

The Stars Looked Upon Us

I stood outside on the west end of the convenience store looking up at the sky. Each star twinkled as I felt the chill of moisture course down my face. My hands and fingertips were cold, not because it was cold but because my blood has been drained inside the pressure of my grip. I was at a dead stop —

I COULD SMELL the exhaust of irreparable cars parked at the front of the store. Although Santa Fe was a wealthy city, there were many less fortunate citizens. Because of the rise in real estate prices, some residents could not afford to maintain their cars or pay their rent. Their powerlessness qualified many parts of the city as a slum.

The orange lights on the side of the building trembled, matching the colors of the city's buildings in the daylight. This light was less comforting than daylight. In the distance the city hummed in desperation, overcome by the shadow of poverty and oppression pushing people to horrible violence. Cars drove around the city with tinted windows, too dark to reveal the occupants. My body's inner organs vibrated with the thump of lowriders and humming lamps. The ocher dusk invited the invasive and elongated pulses.

We stopped at a convenience store for cigarettes. As my friends went in, I lit a cigarette and leaned against the warm, prickly wall. The echo of cars and the buzz of the flickering orange light preoccupied the wavering whispers of ghosts coming from the shadows. My friends finally came out. Dwayne wandered behind the store and disappeared into the blackness. Lane paced in circles behind me as I took a drag from my cigarette. I pinched the butt between my fingers and flicked it into the parking lot. The grunt of an old punctured, piss-yellow, rust-covered pickup truck pulled behind the building. The truck sat up high on portly grey tires and did not move while idling. The penumbra draped the contours of metal and skin. As I turned to tell Lane to walk to the front of the store, I heard the southern metal creak of the unoiled driver's side door. The driver walked to the front and made a beeline toward my friend Dwayne. The other younger guy noticed as I followed his friend and hopped out of the vehicle. The younger one shouted from the pick-up, "I got this one!" Lane ran back inside the store and called the police. The driver ran over to Dwayne, grabbed him around the neck, and threw him to the ground. Dwayne screamed as the driver kicked him. The younger guy jumped on my back and hit me. I managed to turn around and pushed him to the ground. He stumbled to his feet and ran back to the pickup. He reached in and returned with a handgun pointed directly at me.

"What are you going to do now, faggot?"

I stood there.

"Common, faggot! Do something! COME ON, DO SOMETHING!"

I took one steady step back. Then off to the side, into the shadow. From afar, I could hear the reprehensible defeat, a breath almost devoid of life, followed by the slightest moan. My hands became a useless clinch. The silent rage building inside my body was released in my eyes. Tears flowed. I looked to the sky and it was a dusky purple. So far away, the stars twinkled. I waited for the gun to go off, staring down into the small blackness. Dwayne stopped screaming, but I still heard a bagged noise inside the silent dark. I thought the worst.

My ears filled with the pumping silence. Over it I could hear voices yelling. The younger occupant looked towards the front of the store and shouted towards the driver. The driver emerged from the darkness. He ran towards the truck. The younger one flew into the truck like a wolf spider and prey, still pointing the gun in my direction. The truck sped out of the parking lot, crossing the street, running over the median, and disappearing into the populated darkness.

I stood on the west end of the convenience store looking up at the sky. There, each star twinkled as I felt the chill of moisture course upon my face. My hands and fingertips were cold, not because it was cold outside, but because my blood has been drained inside the pressure of my grip. I was at a dead stop —

Hand-painted sign on the
siding of a recruiting office.
Photograph © James Speed Hensinger.

LAURA MAHAL

One Person, One Vote

I SMILED AT the man who approached my desk, his untied work boots caked with mud. His open plaid jacket had a bit of stuffing poking out just below his right pocket. The contrast of that texture, a fluffy bunny's tail, attracted my attention to a glint of metal on his belt. Not in the center, where he sported a large silver buckle of an elk's head, but to the right, over his hip.

Probably a cell phone, I thought, dismissing my inquisitiveness as paranoia.

"I'd like to vote," he said.

"Glad to help," I offered. "Would you prefer a paper ballot or would you rather vote by machine?"

"Don't trust those damn machines," he said, spitting as he reached for his wallet.

Did he seriously just spit on the floor of the voting center? I asked myself.

A big globule glistened near his left boot, refracting the fluorescent lighting of the Sunday School room. The church had rented the space to the county through the end of elections for the use of our Early Voting Center. Bible verses still showed faintly through the big sheets of construction paper we'd used to cover the bulletin boards. "Love your enemy." "Be kind to one another." Puffy pastel hearts surrounded the advice to "Be tenderhearted."

A train whistle blew outside, startling me, and, I noticed, a woman in a blue rain jacket who had just finished voting. As she walked past me, I caught a whiff of her body lotion, pomegranate and blackberries.

She smelled like California.

She stepped right into the spit.

I politely asked the man if he had a form of ID.

"I'm an American. I don't need no ID."

I wondered why in the hell he'd been reaching for his wallet if he wasn't planning to provide a form of identification, but whatever. My job was to take care of the customer. Yet I sure wished my supervisor wasn't across the room behind the enormous TSX voting machines.

"Actually, I do have something I want to show you,"

he said, leaning in close enough for me to see the hairs curling from his nose. He tucked his jacket back, long enough for me to see the glint of metal in a holster on his belt.

Not a cell phone.

He propped his elbows in front of me.

"I said I want to vote. And I mean to do so. But you're gonna give me a stack of ballots, then wait here while I vote every single one of them."

My breath stuck in my lungs like I'd swallowed salt-water taffy down the wrong pipe. I cleared my throat, then did it again, louder, hoping my bipartisan partner would get back with her Keurig coffee. Red nametag for Republican, Black for Independent, or Blue like mine — I didn't give a rat's ass who showed up or what their party affiliation. I just wanted some help.

I started buying for time.

"Sir, I may have misunderstood you. I can only distribute one ballot per person. If you'll just give me your name and birthdate, I can pull you up on my computer in 20 seconds."

My gray metal folding chair screeched as I leaned back on two legs, desperate to draw anyone's attention to what was going down at my station.

My second-line supervisor was helping a man with the distinct shake of Parkinson's to unfold his paper ballot.

My Jolly Red partner was stirring nondairy creamer into her coffee, and sneaking bites of the donut she didn't want anyone to notice she was eating. Her back was to me. She was facing a bulletin board that once had addressed the thrill of victory over sin.

I scanned my desk for tools of self-defense. A ball point pen. A cheap plastic stapler. A vase of black-eyed Susans and a potted chrysanthemum. The jeweled and studded purse of the woman in the blue raincoat who smelled like blackberries and pomegranates and had spit on her expensive shoes.

The man reached for his hip, grunting, "I don't think you understood me."

But I was quicker.

I sprung up and swung my chair like a baseball bat, smashing it as hard as I could onto his head and neck. I vaulted over the table and onto his chest before he could say, *"What the . . .?"*, pinning his arms and knocking his weapon out of the holster. It went spinning across the room toward the provisional ballots table.

I dug my knee into his ribs, right into the white fluff

of the bunny tail on his plaid jacket.

"I understood you just fine," I said.

People ran toward us from every direction. My partner dropped her donut. My supervisor shouted into her cell phone and the second-line supervisor herded voters out of the room. But my attention narrowed to the man at my mercy.

"Don't ever make the mistake of assuming you can intimidate a woman. I was an MP long before I became an elections judge," I said.

"One person, one vote. That's democracy."

Notes on Contributors

CRISOSTO APACHE, originally from Mescalero, New Mexico, on the Mescalero Apache Reservation, currently lives in Denver, Colorado, with his spouse of 17 years. He is Mescalero Apache, Chiricahua Apache, and Diné (Navajo) of the 'Áshįįhí (Salt Clan) born for the Kinyaa'áanii (Towering House Clan). He holds an MFA from the Institute of American Indian Arts in Santa Fe, New Mexico. He teaches writing at various colleges in the Denver area and continues his advocacy work for the Native American LGBTQ/two spirit identity. Crisosto's debut collection ~~GENESIS~~ (Lost Alphabet, 2018) stems from the vestiges of memory and cultural identity of a self-emergence as language, body, and cosmology.

MARYANNA WIENBROEER BRUNKHORST is a returned Colorado native currently residing in Denver, Colorado. She spent seven years in the U.S. Air Force as a C-130 navigator before getting out to go back to school for teaching. She earned a master's degree in TESOL from Middlebury Institute of International Studies at Monterey, and teaches English language courses at the University of Denver's English Language Center. She is also the owner of Chinook Writing and Journaling and facilitates creative journaling workshops throughout the Colorado Front Range area. She is happy to be back in Colorado near her parents, sister, and nieces.

CATH COLWELL lives and writes in Denver, Colorado. She served in the United States Air Force in the Vietnam era where she was the first woman to be assigned to an Intercontinental Ballistic Missile site with secret clearance. She attended the University of Colorado in Boulder, studying social sciences. Currently, she takes care of a parent, and dabbles in real estate.

ALICIA R. CHRISTOPHER was born and raised in a small town in Mississippi. After graduating from high school, she decided to join the U.S. Army. She has always been fascinated with poetry and she has been inspired by great poets like Walt Whitman, Nikki Giovanni, and Maya Angelou. After completing her service in the military, Christopher dedicated her life to educating youth, raising her beautiful daughters, and travelling around the world.

CAROL COVINGTON is the daughter of an Air Force veteran, and she a veteran of the U.S. Army. A recent graduate of the Community College of Denver, Covington is currently enrolled in the summer semester at Metropolitan State University of Denver, where she is majoring in English with a focus on creative writing. Born in Lakehurst, New Jersey, she spent most of her childhood in Wiesbaden, Germany and intermittent tours with her family at Vandenberg Air Force Base in California. Covington began writing in childhood and continued writing greeting cards and poetry. She reads locally in Denver at the Inkwell, Prodigy Coffeehouse, and the iconic Mercury Cafe and is currently working on her first chapbook.

JASON DAVIES is a veteran living, working, writing, and raising a son in Colorado. Building it so they will come.

JAMES SPEED HENSINGER earned a BS in geology and petroleum engineering, an MS in geology (incomplete), and, while serving as an EMT on the Pima County Search & Rescue Team, an MS in library science. He served in the 173rd Airborne Brigade in Vietnam, which has seriously compromised his ability to enjoy time spent in the great outdoors. He has supported himself as a silversmith, Volkswagen mechanic, landscape worker, graduate student, farm laborer, librarian, hardware store clerk, petroleum geologist, handyman, road laborer, EMT, backpacking guide, soldier, night watchman in a funeral home, software engineer, fur hunter, author, photographer, and dishwasher.

ALEXANDRA JACKSON is a recent graduate from the University of Denver with a BA in psychology and creative writing and a MSW. She holds distinction in writing for her senior thesis, "Angels in My Living Room." She hopes to bridge her MSW with her passion for creative writing to increase self-expression and self-esteem while breaking down systematic barriers for marginalized communities. Her works look at the storage of trauma in the biopsychosocial body and the release through anti-oppressive work and counter-narratives.

DAVID LENNON is a 78-year-old USMC vet. He spent 1958–61 doing a tour of duty: Westpac, Okinawa, Formosa (Taiwan), and Japan. He lives in Denver, Colorado.

LAURA MAHAL is a two-time winner of the Hecla Award for Speculative Fiction—for (S)he and WC in 2016 and The Kid from the Other Side in 2017. Her primary focus is literary fiction, but she "dabbles" in other genres, occasionally tapping into her U.S. Army experiences. Mahal's poetry is featured in *Sunrise Summits*, a finalist for the 2017 Colorado Book Awards. Her fiction recently appeared in *Fish*, earning her the chance to read at the West Cork Literary Festival in Ireland. She is the member liaison for Northern Colorado Writers and blogs for The Writing Bug at writingbugncw.com.

LUKE ANTHONY ALFONSO MARTÍNEZ was born in central New México. Growing up in a large, loving family in Durango, Colorado, he began writing poetry as a young boy. In January 1989, after two enlistments and an honorable discharge, Martínez entered the Theatine Seminary in Denver, Colorado, where he professed vows as a Theatine brother. After leaving the seminary, he graduated from the University of Colorado, Denver in 2003 with a Spanish major and a writing minor. He enjoys traveling with his partner Elmer but will always have a love for Colorado and his native New México, the Land of Enchantment. He works for a counseling agency near downtown Denver.

MICHAEL MCANDREW is a veteran of the U.S. Navy. He does most of his writing at his job, working overnight in a residential treatment facility for children with severe trauma. He lives in Denver, Colorado.

KRISTINE OTERO served in the U.S. Army on active duty from 2003–2007, completing two combat tours in Iraq. She went on to complete two years in the Texas Army National Guard before being honorably discharged in 2010. Otero earned an undergraduate degree in psychology from Liberty University and is working on her master's degree in public administration through American Military University. She has been published in 0-Dark-Thirty, and is currently working on a memoir of her military experience. She lives in Denver, Colorado.

Afterword

Under the leadership of former Chairman William Adams, a Vietnam veteran and a philosophy scholar, the National Endowment for the Humanities issued a call to the state humanities councils "to explore the experience of war through the lens of the humanities" through an initiative called "Standing Together." We quickly discovered that the greatest need we could address was to help veterans integrate in their home communities when they return from deployment.

We asked Colorado veterans how we could help them and they let us know that the greatest avoidable difficulty is an inability to ever "come home" from conflict. They suggested a writing workshop and we dreamed of publication of veterans' work and staging readings and conversations about local veterans' writing between veterans and civilians. Because of the preparation prompted by NEH's initiative, we were ready to say "yes" when Denver Veterans Writing Workshop co-leader Jason Arment called and asked for help starting the workshops for veterans. Jason Arment and co-leader Steven Dunn continued to offer the workshops without renumeration until Arts in Society awarded this project with a grant to support the workshops through the fall, winter, and spring of 2017–18 and fund the publication you hold in your hands.

Intended as a reader to spark conversation in Denver-metro communities, Arment, Dunn, and co-editor Bethany Strout selected the writing and worked with the authors, some new to writing, to prepare their work for publication.

Colorado Humanities is proud of the work collected here and, as always, we look forward to the conversation!

think. learn.
join the conversation.

This book is intended for adult readers. The views, findings, conclusions, or recommendations expressed in this publication do not necessarily represent those of the National Endowment for the Humanities, Colorado Humanities, or Arts in Society funders.